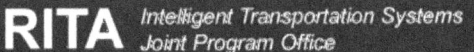 **RITA** *Intelligent Transportation Systems Joint Program Office*

An Approach to Communications Security for a Communications Data Delivery System for V2V/V2I Safety:

Technical Description and Identification of Policy and Institutional Issues

www.its.dot.gov/index.htm
White Paper — November 2011

FHWA-JPO-11-130

U.S. Department of Transportation

Research and Innovative Technology Administration

Produced by Crash Avoidance Metrics Partnership and the John A. Volpe National Transportation Systems Center

ITS Joint Program Office
Research and Innovative Technology Administration
U.S. Department of Transportation

Notice

This document is disseminated under the sponsorship of the Department of Transportation in the interest of information exchange. The United States Government assumes no liability for its contents or use thereof.

Technical Report Documentation Page

1. Report No. FHWA-JPO-11-130	2. Government Accession No.	3. Recipient's Catalog No.
4. Title and Subtitle An Approach to Communications Security for a Communications Data Delivery System for V2V/V2I Safety: Technical Description and Identification of Policy and Institutional Issues		5. Report Date November 2011
		6. Performing Organization Code
7. Author(s) The Crash Avoidance Metrics Partnership (CAMP) and the VIIC From the Volpe Center: Anita Kim, Valerie Kniss, Gary Ritter, Suzanne M. Sloan		8. Performing Organization Report No.
9. Performing Organization Name And Address U.S. Department of Transportation Research and Innovative Technology Administration John A. Volpe National Transportation Systems Center Cambridge, MA 02142		10. Work Unit No. (TRAIS)
		11. Contract or Grant No.
12. Sponsoring Agency Name and Address Intelligent Transportation Systems (ITS) Joint Program Office (JPO) Research and Innovative Technology Administration U.S. Department of Transportation 1200 New Jersey Ave., S.E. Washington, D.C. 20590		13. Type of Report and Period Covered Final Report, 2011
		14. Sponsoring Agency Code
15. Supplementary Notes		
16. Abstract This report identifies the security approach associated with a communications data delivery system that supports vehicle-to-vehicle (V2V) and vehicle-to-infrastructure (V2I) communications. The report describes the risks associated with communications security and identifies approaches for addressing those risks. It also identifies and describes the policy and institutional issues that require focus in support of implementation and operations, as well as the balance needed among the priorities of security and safety with cost, privacy, enforcement, and other institutional issues. The approach described in this report is a first step in identifying the technical and policy requirements that will form the basis for a prototype model that will be tested during the 2012-2013 Safety Pilot Model Deployment, located in Ann Arbor, Michigan. The prototype will be tested along with draft policies and procedures. Results of the test will inform the requirements, specifications, and guidelines for implementing an operational system.		
17. Key Words Intelligent Transportation Systems (ITS), Joint Program Office (JPO), Connected Vehicle, V2V, V2I, privacy, security, public key infrastructure (PKI), physical security elements, communications channel capacity, communications media, communications security risks, enforcement techniques, public safety, legal deterrence, governance, risk mitigation, certificate management entity, certificate authority, registry authority, communications platform, slander attacks, framing attacks, Sybil attacks, denial-of-service, message linking, software manipulation, sensor manipulation		18. Distribution Statement This document is available to the public through the National Technical Information Service, Springfield, Virginia 22161.
19. Security Classif. (of this report) Unclassified	20. Security Classif. (of this page) Unclassified	21. No. of Pages 52 22. Price N/A

Form DOT F 1700.7 (8-72)   Reproduction of completed page authorized

Table of Contents

Executive Summary 5

Section I: Approach to Communications Security for a Communications Data Delivery System for V2V/V2I Safety 11

    I.A Objectives and Requirements 12

    I.B Configuration and Design 13

        I.B.1. Public Key Infrastructure 14

        I.B.2. Technical Solutions 16

        I.B.3. Policy Options 16

        I.B.4. Summary 17

Section II: Security Risks/Threats 18

    II.A Attacks on the User 18

    II.B Attacks on the Communications System 22

    II.C Summary 24

Section III: Addressing Security Risks—Technical and Policy Options 26

Section IV. Preliminary Policy Analysis 36

    IV.A Organizational and Operational Models for a Certificate Management Entity 36

    IV.B Legal Deterrence and Enforcement 37

    IV.C Privacy 38

    IV.D System Implementation: Costs and Sustainable Funding/Financing/Investments ... 38

    IV.E Governance 39

Section V. Conclusion 40

Appendix A: Details on Policy Research 41

    A.1 Policy Research for Certificate Management Entities: 41

    A.2 Policy Research for Communications Data Delivery System Options: 42

List of Figures

Figure 1: Key Elements of Proposed Security Approach 14
Figure 2: Proposed PKI Approach to Communications Security 15

List of Tables

Table 1: Assessment of Safety Risk through Attacks on the User 21
Table 2: Assessment of Privacy Risk through Attacks on Communications Infrastructure 24

Executive Summary

For 2010-2014, the primary focus of the United States Department of Transportation's (USDOT) Intelligent Transportation System (ITS) Program is a multimodal research initiative[1] focused on developing rapid and accurate wireless communication and data exchange among vehicles, roadside equipment (RSE), and passengers' personal communications devices. This innovative use of wireless communications offers an unprecedented opportunity to create an information-rich, connected vehicle environment that may transform surface transportation safety, mobility, and environmental performance.

The connected vehicle environment will use wireless, short-range communications to deliver data and create a dynamic data exchange between and among vehicle-to-vehicle (V2V), vehicle-to-infrastructure (V2I), and vehicle–to-mobile device (V2D). This exchange will support a variety of cooperative applications and systems.[2] Crash-avoidance safety applications are the highest priority and they establish the **minimum acceptable technical and policy requirements for security and trust** for a communications data delivery system. An additional important requirement is user acceptability which is based not only on the level of security but also on an appropriate balance between privacy, cost, and safety, among other important factors.

In considering technical and policy requirements as well as other requirements, it is worth noting that the envisioned communications data delivery system is establishing new ground as a system supporting safety-of-life using wireless communications. To date, wireless communications systems tend not to support safety-critical functions; nor do safety-critical systems tend to be based upon wireless systems. As such, the approach to security of the communications data delivery system draws from industry best practices but also establishes new practices to meet the specific needs and requirements of a connected vehicle environment.

Presented in this document is the proposed approach to communications security for a V2V/V2I safety communications data delivery system ("V2V/V2I communications system"). The objectives of this document are the following:

- Identify the **technical, policy, and institutional requirements for communications security** (based on USDOT, industry, and stakeholder needs);
- Describe the most probable **user and system risks**, their types and severity;
- Examine the varying levels of **security options available to address the risks**; and

---

[1] The research is administered through the U.S. Department of Transportation's (US DOT) ITS Joint Program Office (ITS JPO) and conducted in partnership with the Department's surface transportation modal administrations. The five year research agenda is described in the **ITS Strategic Research Plan, 2010-2014** (http://www.its.dot.gov/strategic_plan2010_2014/index.htm).

[2] More detailed information on the transformative nature of these cooperative systems can be found in ITS Strategic Research Plan, 2010-2014 (http://www.its.dot.gov/strategic_plan2010_2014/index.htm and in the report, Frequency of Target Crashes for IntelliDrive Safety Systems at: http://www.nhtsa.gov/DOT/NHTSA/NVS/Crash%20Avoidance/Technical%20Publications/2010/811381.pdf.

- Examine the **policy and institutional issues associated with this approach and resulting impacts on safety, privacy, user acceptance, and cost**. It is expected that the most effective approach to communications security will leverage both technical and policy measures to optimally and effectively address security requirements.

ES.A   Approach to Communications Security—Requirements

The approach to communications security (herein, "the approach") was developed in partnership with industry[3] and with input from five teams of security experts whose expertise is based on experiences studying, designing, and implementing leading-edge security options for a variety of industries.[4] The approach draws upon globally-accepted best practices, but further configures these practices to meet the unique requirements necessary for a connected vehicle environment and, specifically, crash-avoidance safety applications. Four high-level, critical requirements form the foundation of the approach:

- *Protection of Privacy*: The communications security system shall not allow for identification of a person through personally-identifiable information (PII) within messaging contents.
- *Secure Communications*: All communications transmitted and received from a vehicle shall be secure. This includes both one-way and two-way communications. Messages will support delivery and management of security credentials and will be encrypted to prevent eavesdropping and tampering over the communication channel.
- *Trusted Communications*: All communications exchanged between vehicles shall be trusted. Trust will be established through a user authentication process, which determines permissions and allowed actions with the system and other users.
- *Scalability to Enable Nationwide Adoption*: The security approach shall be scalable to support a population of over 250 million vehicles using the system.

In this approach, messages from an RSE are considered secure as the RSE will receive and use digital certificates to secure transmitted information. Other aspects of V2I security, however, were not included because they require further research. For example, hardware access and other elements of physical security for RSEs and aftermarket devices (ASDs) will require additional analysis, in particular as the system expands to include V2I mobility and environmental applications. At this point, it is expected that they will leverage similar concepts and elements from the security approach for V2V/V2I safety.

---

[3] The Crash Avoidance Metrics Partnership (CAMP) provides an OEM-oriented research consortium under which various stakeholders can collaborate as desired on pre-competitive crash avoidance research projects of mutual interest. The VSC3 Consortium, consisting of Ford, General Motors, Honda, Hyundai/Kia, Mercedes, Nissan, Toyota, and Volkswagen/Audi, was formed under the CAMP agreement to conduct pre-competitive research on 5.9 GHz DSRC cooperative safety technologies and applications.

[4] The security teams are as follows: Carnegie Mellon University; the University of Illinois at Urbana-Champaign and Telcordia; GM India Science Labs; Security Innovation and eScrypt.

ES.B   Security Risks

There are two broad categories of risk associated with communications systems:

- **Attacks on the user/risks to safety and user acceptance:**   Attacks on the user are aimed at directly impacting the safety of users and indirectly impacting system acceptance. With these types of attacks, attackers have two goals in mind: (1) to cause users to make bad driving decisions resulting in an accident, congestion, or reroute of a driver; and (2) to reduce users' faith in the system as messages become unreliable or unavailable.

- **Attacks on the communications system/risks to privacy:** Attacks on the communications system are attacks that lead to threats to privacy or cause drivers to bear extra administrative or legal burdens within the system. These types of attacks occur in two categories: (1) privacy attacks through tracking the location or driving route of a particular person; and (2) slander or framing attacks which involves the false reporting of misbehavior from a vehicle, resulting in an otherwise valid driver being removed from the system.

Section II of this document will discuss in greater detail the expert evaluation of potential threats to the V2V/V2I communications system, an assessment of the probability of such attacks, and an analysis of the level of risk resulting from the threat. The overall conclusion of the expert analysis is that there are a low number of high risk attacks on the user and there appears to be minimal risk to safety in the event of a successful attack. This is, in part, due to the short-range nature of the communications and thus the geographic limitations to such attacks. Instead, analysis concludes that the greater risk is in reducing acceptance of the system to the point that users ignore it.

With regard to attacks on the system, the main risk appears to be related to privacy. To minimize this risk, the operational system is expected to use randomly assigned identifies that contain no identifying information. This configuration would make it necessary for an attacker to have some combination of inside knowledge of the system, physical access to the vehicle, or sizable investment resources to launch a successful attack on privacy. The complexity of such an attack is expected to deter attackers, reducing potential threats and ensuring that the risk to privacy is no greater than in today's world where attackers utilize more accessible methods such as cell phone tracking or physically following a vehicle.

ES.C   Approach to Communications Security—Configuration of Technical and Policy Options

Section III describes a set of technical and policy options designed to reduce risks specific to a V2V/V2I communications environment. The proposed approach combines three elements as the basis for a robust security solution:
1. Public Key Infrastructure (PKI)
2. Technical solutions (vehicle, hardware, and software)
3. Policy options

PKI is an umbrella term used to describe the hardware, software, people, policies, and procedures needed to create, manage, store, distribute, and revoke digital certificates. PKI forms the foundation of the approach to V2V/V2I communications security by employing *public key cryptography* to authenticate a sender or encrypt data, thereby producing trusted and secure messages.

In addition to PKI, the approach integrates additional technical design elements (measures for providing security at both the vehicle and system level) and implementation of enforcement policies to deter attacks. The approach also employs techniques to detect misbehavior and remove the misbehaving entity. In areas where security risks cannot be addressed through technical design alone, policy mechanisms such as governance, legal deterrence, and enforcement can serve a critical supporting role.

ES.D   Policy Issues Requiring Further Research

Section IV of this report summarizes the key policy and institutional issues that will need to be addressed in support of an operational system. The text additionally notes some of the inherent conflicts that may require decision makers and stakeholders to make choices or balance priorities.

The policy and institutional issues that are considered most significant (and thus will result in additional policy research) include:

- Analyses on how to most effectively design **organizational and operational entities** that will support security credential (certificate) management, legal deterrence, misbehavior detection, and revocation. Outstanding policy and institutional issues include questions on cost, whether to split the entities for enhanced privacy, personnel and equipment needs, and policies and procedures.
- Identification of the specific types of **legal deterrence and enforcement policies** that will act to prevent or mitigate misbehavior within the system. Outstanding policy questions include the determination of authority for enforcement.
- Development of a strategy for updating **and implementing the 2007 privacy principles**, including development of practicable options for putting the principles into use.
- Analysis on **implementation options** that compares different configurations using infrastructure and non-infrastructure options. Further, analysis on the types of **sustainable funding, financing, investment, and/or revenue sources** available with these implementation options that address the needs for funding initial deployment as well as ongoing operations, and maintenance.
- The identification of the level and type of **governance and authorities** required for implementation of the organizational and operational models and with the communications data delivery system.

ES.E An Evolutionary Path for Developing Communications Security

Development Process

Creating an approach represents the first step in the development process of designing and implementing a secure communications data delivery system for V2V/V2I Safety. The development process uses an incremental path that allows for review and decision making at critical junctures throughout research and development before a full system is deployed. This type of process encourages stakeholder input and provides opportunity for continuous refinement. The steps of the development process are defined below.

- **Approach:** The approach represents the first step in identifying industry best practices and tailoring these practices to meet the requirements of a V2V/V2I environment for preventing, detecting, and mitigating security risks. The approach is the main focus of this document.

- **Design:** The design is the second step in the development process that will structure the requirements and technical elements into a representative prototype (due in Spring 2012). The design will assist in the targeted development of the policy, institutional, and organizational elements that support the operations of the system.

- **Model:** In this step, the technical prototype is combined with organizational and operational elements, resulting in a model system for testing and evaluation in a real world environment. The testing will occur during Safety Pilot Model Deployment (see side textbox for description; results from testing due in Summer 2013).

- **System:** The final step is to analyze the test results of the prototype model and to develop final technical and institutional requirements, specifications, objective test procedures, and policies for an operational system (due in early 2014).

Further Research

The approach to communications security detailed in this document is configured to support successful initial implementation of crash-avoidance safety applications. From a technical perspective, it is possible for the approach to be scaled and expanded to support a wider range of safety, mobility, and environmental applications (e.g. notification of school zones, tolling, or fuel efficiency), although such an expansion is likely to require further technical, policy, and institutional research.[5] As noted previously, further research is also needed to address security for other elements that are external to the communications data delivery system. These elements include:

- **Security of devices:** mobile (personal) device security and aftermarket / retrofit device security;

- **Security of infrastructure:** security of communications infrastructure nodes (or roadside equipment), although it is not yet clear how much infrastructure will be needed; and

---

[5] From a technical perspective, this expansion will likely require further research on the integration of multiple communications platforms as well as the communications security of the data delivery from users and/or infrastructure nodes to back office systems. From a policy perspective, this expansion will require further research on privacy and governance.

- **Organizational security:** security of the entity(ies) that manages the security credentials (referred to as a Certificate Authority, or CA) and associated policies for access to data and the system.

Although still conceptual, developing an approach at this time serves a number of important purposes. First, it allows all stakeholders involved to:
- Analyze the strengths, limitations, and vulnerabilities of the proposed approach
- Discuss the policy and institutional priorities and trade-offs; and
- Identify potential obstacles or challenges to implementation and operations.

Second, it allows stakeholders to provide feedback on how the approach supports their needs. Their inputs will be provided to the technical teams as the approach moves into a preliminary design and prototype. It also provides an opportunity to identify whether there are alternative approaches that may be more appropriate or effective.

Last, the development of the approach has highlighted the need to conduct policy research on key institutional elements that support implementation of an operational system. As noted above in section ES.D, research that is planned or underway includes:
- Development of options and definitions of proposed roles and responsibilities for **organizational and operational management entities**. This includes identification of enforcement techniques and governance needs that meet key requirements, such as privacy protection.
- Development of **viable financial models** to support initial implementation as well as ongoing operations and maintenance needs of the communications data delivery system.

ES.F   Organization of the White Paper

To present the proposed approach to communications security, this document is organized as follows:
- Section I: A description of the proposed approach.
- Section II: Analysis of the types of risks associated with communications security.
- Section III: An understanding of how the proposed approach is configured to address the identified risks. This section highlights the role of policy in combination with technical solutions.
- Section IV: Identification of the policy and institutional trade-offs. A preliminary trade-off analysis has resulted in an ability to generally consider security requirements vis-à-vis other priorities such as: cost, impact to safety, the level of institutional oversight and management, whether any new authorities are required for this approach, and enforcement needs.
- Section V: A description of the technical or policy gaps that still need to be resolved as the approach moves through the process of setting requirements, designing and developing a working prototype, testing, and implementing the system.

Section I: Approach to Communications Security for a Communications Data Delivery System for V2V/V2I Safety

For 2010-2014, the primary focus of the United States Department of Transportation's (USDOT) Intelligent Transportation System (ITS) Program is a multimodal research initiative[6] focused on developing rapid, and accurate wireless communication and data exchange among vehicles, roadside infrastructure, and passengers' personal communications devices. This innovative use of wireless communications offers an unprecedented opportunity to create an information-rich, connected environment for transportation that may transform surface transportation safety, mobility, and environmental performance.

The connected vehicle environment will use wireless, short-range communications to deliver data and create a dynamic data exchange from vehicle-to-vehicle (V2V), vehicle-to-infrastructure (V2I), and vehicle–to–mobile device (V2D). This exchange will support a variety of cooperative applications and systems.[7] Crash-avoidance safety applications are the highest priority and they establish the **minimum acceptable technical and policy requirements for security and trust** associated with the communications data delivery system. User acceptability is another priority and reflects the balance between the required level of security, privacy, cost, and safety. Acceptability, among other important factors, will form an important basis for adoption and use of the system and its safety applications.

In considering technical and policy requirements as well as other requirements, it is worth noting that the envisioned communications data delivery system is establishing new ground as a safety-critical system using wireless communications. To date, wireless communications systems tend not to support safety-critical functions; nor do safety-critical systems tend to be based upon wireless systems. As such, the approach to security of the communications data delivery system draws from industry best practices but also establishes new practices to meet the specific needs and requirements of a connected vehicle environment.

Presented in this document is the proposed approach to communications security for a V2V/V2I safety communications data delivery system. The objectives of this document are the following:

---

[6] The research is administered through the U.S. Department of Transportation's (US DOT) ITS Joint Program Office (ITS JPO) and conducted in partnership with the Department's surface transportation modal administrations. The five year research agenda is described in the ITS Strategic Research Plan, 2010-2014 (http://www.its.dot.gov/strategic_plan2010_2014/index.htm).

[7] More detailed information on the transformative nature of these cooperative systems can be found in ITS Strategic Research Plan, 2010-2014 (http://www.its.dot.gov/strategic_plan2010_2014/index.htm and in the report, Frequency of Target Crashes for IntelliDrive Safety Systems at: http://www.nhtsa.gov/DOT/NHTSA/NVS/Crash%20Avoidance/Technical%20Publications/2010/811381.pdf.

- Identify the technical, policy, and institutional requirements for providing communications security (based on USDOT, industry and stakeholder needs);
- Describe the most probable user and system risks, their types and severity;
- Examine the varying levels of security options available to address the risks; and
- Examine the policy and institutional issues associated with this approach and resulting impacts on safety, privacy, user acceptance, and cost. It is expected that the most effective approach to communications security will leverage both technical and policy measures to optimally and effectively address security requirements.

I.A   Objectives and Requirements

The approach to communications security (herein, "the approach") was developed in partnership with industry[8] and with input from five teams of security experts whose expertise is based on experiences studying, designing, and implementing leading-edge security options for a variety of industries.[9] The teams drew upon industry best practices, but adapted these practices to meet the unique requirements necessary for providing V2V/V2I safety applications.

In developing the approach, the analysis focused generally on two key questions —*What risks (security) are present and how can these risks be addressed? What technical and policy elements are required for a comprehensive communications security system?* In answering these questions, the team began to identify several issues influencing development of the approach, such as:

- Range of likely attacks and resulting risks to privacy, safety, and user acceptance of the system;
- Technical, policy, and legal options available to deter attacks;
- Processes for misbehavior detection, revocation mechanisms, and potential enforcement policies;
- Communication channel requirements for supporting secure data exchange between users and the system; and
- Options for establishing and maintaining trust through authentication, non-refutability, traceability, and auditing.

In partnership with stakeholders, the USDOT defined four, high-level but critical requirements that the proposed approach must meet in support of all of the V2V safety applications and a subset of the V2I safety messages. These requirements are necessary to ensure acceptance of the communications data delivery system. Safety of

---

[8] The Crash Avoidance Metrics Partnership (CAMP) provides an OEM-oriented research consortium under which various stakeholders can collaborate as desired on pre-competitive crash avoidance research projects of mutual interest. The VSC3 Consortium, consisting of Ford, General Motors, Honda, Hyundai/Kia, Mercedes, Nissan, Toyota, and Volkswagen/Audi, was formed under the CAMP agreement to conduct pre-competitive research on 5.9 GHz DSRC cooperative safety technologies and applications.

[9] The security teams are as follows: Carnegie Mellon University; the University of Illinois at Urbana-Champaign and Telcordia; GM India Science Labs; Security Innovation and eScrypt.

the user (and thus, security of the data and messages) was given the highest priority, with protection of privacy considered as a significant requirement for user adoption[10].

The critical requirements include:

- **Protection of Privacy**: The communications security system shall not allow for identification of a person through personally-identifiable information (PII) within messaging contents.

- **Secure Communications**: All communications transmitted and received from a vehicle shall be secure. This includes both one-way and two-way communications. Messages will support delivery and management of security credentials and will be encrypted to prevent eavesdropping and tampering over the communication channel.

- **Trusted Communications**: All communications exchanged between vehicles shall be trusted. Trust will be established through a user authentication process, which determines permissions and allowed actions with the system and other users.

- **Scalability to Enable Nationwide Adoption** The security approach shall be scalable to support a population of over 250 million vehicles using the system

In this approach, messages from an RSE are considered secure as the RSE will receive and use digital certificates to secure transmitted information. Other aspects of V2I security, however, were not included because they require further research. For example, hardware access and other elements of physical security for RSEs and aftermarket devices (ASDs) will require additional analysis, in particular as the system expands to include V2I mobility and environmental applications. At this point, it is expected that they will leverage similar concepts and elements from the security approach for V2V/V2I safety.

I.B Configuration and Design

With objectives and requirements identified, the expert teams proposed three elements that provide the basis for the proposed security approach:

- Public Key Infrastructure (PKI) scheme
- Technical solutions (vehicle, hardware, and software security)
- Policy options

---

[10] In this document, a user is defined as all end users or users that send and receive messages through the System.

Figure 1: Key Elements of Proposed Security Approach

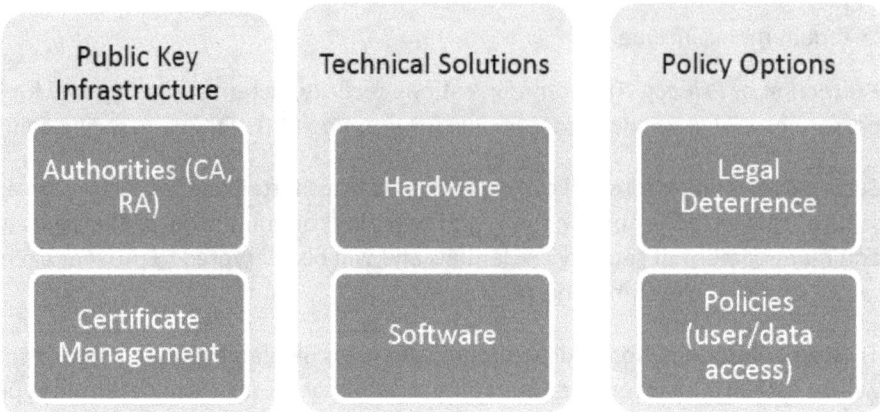

I.B.1. Public Key Infrastructure

PKI is an umbrella term used to describe the hardware, software, people, policies, and procedures needed to create, manage, store, distribute, and revoke digital certificates. As defined by experts, a PKI allows users of an unsecure public network to securely and privately exchange data through the use of a public and a private cryptographic key pair that is obtained and shared through a trusted authority. Although the components of a PKI are generally the same throughout many practices, a number of different approaches are employed to meet specific requirements.[11]

PKI assumes the use of *public key cryptography*, one of the most common methods for authenticating a message sender or encrypting a message. From an institutional perspective, a PKI typically consists of the following:

- A **certificate authority (CA)** that issues and verifies **digital certificates**. A certificate includes the public key or information about the public key;
- A **registration authority (RA)** that acts as the verifier for the certificate authority before a digital certificate is issued to a requestor;
- One or more **directories** where the certificates (with their public keys) are held; and
- A **certificate distribution and management system** which includes the communications system and its organizational and operational elements.

For the connected vehicle environment, PKI will be the basis for security and will be configured to provide a level of reliability, sensitivity, and redundancy needed for crash-avoidance safety applications. The primary function of the PKI is to allow users to exchange data through a trusted authority using authentication credentials (certificates). In addition to establishing trusted messages, the PKI will provide users with authorization credentials and facilitate certificate revocation in the event of misbehavior. While this framework for trusted communications meets the requirements for crash-avoidance safety, it is capable of being expanded

---

[11] Definitions located at: http://searchsecurity.techtarget.com/definition/PKI ; http://en.wikipedia.org/wiki/Public_key_infrastructure; and in Cryptography Decrypted by H.X. Mel and Doris Baker, Addison-Wesley, December 2000.

to support mobility and environmental applications (e.g. tolling or parking reservations). Adapting this framework to fit non-safety applications will require further technical, policy, and institutional research.

Figure 2 illustrates the approach using basic PKI principles. It displays use of a certificate authority and a certificate distribution and management system for creating trusted messages between vehicles. Details of different PKI components and their application in the approach will be discussed in Section III.

Figure 2: Proposed PKI Approach to Communications Security

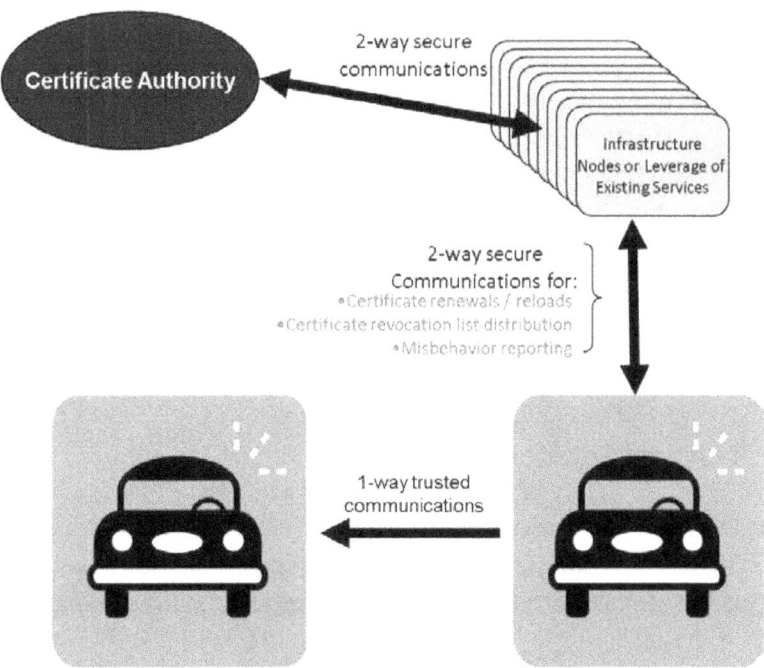

- A one-way *trusted* communications between vehicles (V2V) supports the broadcast of a Basic Safety Message (BSM) once every tenth of a second, which will be received by any neighboring vehicles within range[12]. The BSM is used to exchange "vehicle state" data[13] for use in safety applications; thus, technical requirements include high latency, accuracy, reliability, and speed necessary for crash-avoidance safety applications to establish nearly immediate communications with other vehicles. The messages broadcast by vehicles are trusted, but not encrypted due to the time it would take to encrypt and unencrypt each message between vehicles, potentially hindering the immediate response times required for safety applications. This approach is the

---

[12] The required range is a 300 meter radius.
[13] The BSM is described by the SAE J2735 standard, part 1 which is likely to include a vehicle's temporary ID (instead of containing PII), time, position latitude, longitude, elevation, vehicle speed, transmission state, heading, steering wheel angles, acceleration, yaw rate, and break system status. It will also include data on a vehicle's path and current vehicle dynamics. Most data is measured by a vehicle's sensors. Time and location are derived from the vehicle's GPS signal. The final requirements will depend on the results of scalability testing at a later date. For more information on J2735, see: http://www.sae.org/standardsdev/dsrc/DSRCImplementationGuide.pdf.

same for all safety messages that are transmitted to and from all devices, including safety messages from RSEs and ASDs.

- A two-way *trusted and secure* <u>communications between a vehicle and a certifying authority</u> may or may not require an infrastructure component. Two-way secure communications are encrypted and support the issuing of certificates and the certificate revocation list (CRL) which alerts users of misbehaving actors. These communications also support the reporting of misbehavior occurring within the system. At full deployment, it is expected that each vehicle will communicate with the CA at least once a day.

Under the approach, each new vehicle may have on-board equipment (OBE), which includes an on-board unit (e.g. computer module), display and DSRC radio. This equipment provides an interface to vehicular sensors, as well as a wireless communications interface to the CA. The on-board unit will store the vehicle's certificates which will be encrypted and grouped in batches. Certificates will only able to be decrypted, or unlocked, with keys that are issued from the CA.

Options for the type of communications channel that will support the approach include dedicated short-range communications (DSRC), cellular or Wi-Fi, among others. Communications will go from the vehicle through an infrastructure node or use an already existing system to communicate with the CA. Ongoing policy research will analyze channel requirements as well as various business models capable of supporting channel use.

I.B.2. Technical Solutions

While PKI provides a framework for security, other technical elements can be incorporated at the local (vehicle) level to prevent misbehavior and detect misbehavior already occurring within the system. These technical solutions should work in combination with PKI and policies to make up the comprehensive security approach.

Security at the local level can be integrated into two different areas:
- Hardware: Includes such controls as a standard controller on the vehicle and tamper-proof encasements, among others; and
- Software: Includes functionality checks and misbehavior detection processes.

Hardware security supports restricting vehicle access and vehicle protection from tampering. Software security includes regular checks to identify misbehavior. Similar technical solutions will have to be applied for roadside equipment and aftermarket safety devices also. Further research on the specifics of RSE and ASD application is yet to be conducted.

I.B.3. Policy Options

Policies are an important component of the security approach. They provide non-technical solutions for deterring and addressing misbehavior. Such policies can include:
- Legal deterrence for physical tampering with vehicle's on-board equipment;
- Legal options for preventing misbehavior beyond revocation from the system;
- User access policies; and
- Policies that split the certificate management entity(ies) and ensure that accessing full information about any user will require the authority to gain access to multiple systems and sources.

I.B.4. Summary

Each of the key elements of the security approach (PKI, technical solutions, and policy options) is described in greater detail in later sections of this document. Similar elements must also be applied to roadside equipment and aftermarket devices in order to develop a comprehensive security approach. Further research on implementing a security approach for RSEs and ASDs is underway. In identifying the level of security needed for each element, an analysis of potential risks and threats was performed. These risks and threats are discussed in the following section.

Section II: Security Risks/Threats

Identifying the type, probability, and potential impact of security risks or threats to the communications system was a critical first step in developing the proposed approach. By understanding the potential threat to the V2V/V2I communications system, an appropriate level of security could be designed into the approach.

The concept of risk can be interpreted in multiple ways. In this evaluation of risk, "system risk" is defined as the product of the probability of a successful attack taking place and its impact on the system. The probability, or likelihood, of a successful attack is correlated to two factors:

- **Overall cost to mount an attack**: Considers financial and time costs. It also accounts for the physical difficulties or complexity in mounting an attack as well as potential benefits.
- **Level of deterrence**: Refers to effectiveness of current laws and penalties if the attacker is caught in preventing or deterring attacks

Impact to the system requires quantifying the chance that an attack will lead to an accident, congestion, or route change, which depends on human response in a particular situation. Since determining human response is difficult, the impact to the system metric is converted into an estimated number of false messages that a driver will encounter. Overall, the aim of the approach is to reduce the number of false messages that a driver will encounter to decrease the level of safety risk. A review of threats to existing communications systems suggests that successful attacks would likely impact the following:

- User safety;
- Personal privacy;
- User acceptance of the system; and
- Communications operations.

Potential attacks on the communications system fall into two general categories: a) attacks on the user and b) attacks on the communications system. This section presents a summary of the expert evaluation of these types of attacks and their associated levels of risk.

II.A Attacks on the User

Attacks on the user are aimed at directly impacting the safety of users and indirectly impacting system acceptance. With these types of attacks, attackers have two goals in mind:

- Cause users to make bad driving decisions resulting in an accident, congestion, or reroute of a driver; and
- Reduce users' faith in the system as messages become unreliable or unavailable.

While there could be additional motives for attackers to mount an attack on the user, these two represent the most common motives.

II.A.1 Attack Methods

To mount an attack on the user, an attacker must send messages via a radio in real-time to neighboring vehicles. These attacks are limited to the range of a radio, which is a 300 meter radius. There are three general methods an attacker can use to initiate an attack on the user:

- **Key Extraction**: Physically removing a vehicle's credentials (certificates and keys) from the on-board unit, and then using these credentials on other DSRC radios to create and distribute seemingly legitimate messages to neighboring vehicles.

- **Software Manipulation**: Installing malicious software on the vehicle's on-board unit to create messages containing arbitrary or altered information. The attacker can also manipulate existing software to extract certificates and keys.

- **Sensor Manipulation**: Interfering with the vehicle's sensor output to alter, inject, or suppress messages that originate from internal vehicle systems; or interfering with the sensor input that directly reports vehicle behavior or external circumstances.

Policy Highlight:
Key extraction poses the most significant risk to the system because the attacker is no longer restricted to attacks within the range of a single radio or vehicle. Key extraction allows an attacker to use multiple radios in multiple geographic locations.

Compromised keys can be used to create false messages, affecting user safety and system acceptance, but can also be used in privacy and framing attacks through attacks on the system Infrastructure.

There are existing laws for vehicle tampering which result in legal consequences. Policy research will explore the relevance of these laws in protecting a vehicle's on board equipment.

The methods of attack noted above result in an attacker being able to create false messages which are then distributed to neighboring vehicles. They also require physical access and manipulation of the vehicle.

Another type of method for launching an attack on the user is **Denial of Service (DoS)** attacks. These attacks result in valid messages being suppressed or not received by the vehicle. These attacks do not require physical access to the vehicle, but do require access through a radio. They include:

- **Denial of Computation:** Sending large amounts of bogus messages[14] to a vehicle to cause the on-board unit to be overwhelmed with processing tasks, resulting in not receiving valid messages.

- **Denial of Communication:** Jamming the wireless band with a sufficiently powerful signal, denying vehicles the opportunity to transmit messages.

All vehicles have sensors that detect or measure a vehicle's movements such as acceleration, wheel movement, etc. The sensors transfer this collected information to the vehicle's internal computer in the form of wireless signals. Similarly, the sensors will transfer this information to the V2V applications as a basis for generating warning messages (V2V applications will not be used to override driver controls). Thus, a DoS attack has the ability to impair the V2V on-board equipment and applications, but not to impair the vehicle's internal computer (or Central Processing Unit, CPU).

---

[14] Bogus messages are defined as messages with an invalid signature or authentication tag.

II.A.3 Risk Level

Barriers exist on vehicles today that make launching a successful attack difficult. The chance of improper vehicle access occurring depends on the effectiveness of the vehicle's anti-theft solution, making some attacks no different from any other attack that can occur if a vehicle (or aftermarket device or piece of roadside equipment) is left unsecured. Thus, the threat of certain user attacks is similar to other more traditional threats (vehicle theft, equipment vandalism, etc.)

Also, an attacker must gain physical access to a vehicle for an extended period of time to launch a successful attack. The technological complexity of vehicle hardware and software provides a level of tamper-resistance. It is likely that only an attacker who is familiar with the vehicular architecture would be able to compromise a key, sensor, or GPS receiver.

It is possible that an attacker could extract keys or manipulate software or sensors on his own vehicle, however, as discussed in later sections, linkability of keys to owners will, in general, encourage attackers to focus on devices owned by other people as targets for key extraction. An attacker is unlikely to use his vehicle in an attack, as this may increase the chances of being caught. An attacker would most likely need to steal a vehicle or collude with a repair garage or rental car agency to gain access to a vehicle's on board unit.

Table 1 on the following page describes the methods for launching attacks on the user and the associated risk[15] to safety relative to an attacker's ability level. The risk levels take into account the theoretical implementation of technical elements recommended in the proposed approach. The risk assessment results in Table 1 do not factor in legal deterrence and enforcement methods that could provide an added layer of security to prevent against attacks on the user. Importantly, Table 1 evaluates risk based on the results of a comparison of the transportation environment with "no V2V/V2I system" versus a "perfect V2V/V2I system".

---

[15] Risk = (impact to the system) x (likelihood of a successful attack).

Table 1: Assessment of Safety Risk through Attacks on the User

Methods of Attack		Requires Physical Access	Level of Risk		
			A1	A2	A3
Key Extraction	Physically entering the vehicle and removing keys from the OBU	✓	Low	Medium	High
Software Manipulation	Use APIs[16] to extract keying material	✓	Low	Low	Medium
	Install software on device to create messages containing arbitrary information	✓	Low	Low	Medium
	Install software on device to alter information (i.e. system clock or sensor inputs)	✓	Low	Low	Medium
Sensor Manipulation	Interfere with input to CAN bus that directly report vehicle behavior (e.g. brakes)	✓	Low	Low	Medium
	Interfere with output from CAN bus to application processor	✓	Low	Medium	Medium
	Interfere with input from sensors that report external circumstances (e.g. GPS, lane marker detectors)		Low	Low	Medium
Denial of Service	Jamming the channel (denial of communication)		Low	Low	Low
	Send false messages that cause true messages to be ignored (denial of computation)		Low	Low	Low

The level of risk is dependent upon the ability of an attacker to carry about a successful attack. Attacker ability levels are as follows:

A1 = Clever Outsider – A talented engineer and/or cryptographer who does not possess any inside knowledge.

A2 = Knowledgeable Insider – An insider who possesses detailed knowledge about the system (security and non-security related) and has access to its specifications.

A3 = Funded Organizations – An organization that has access to substantial resources and furthermore possesses the capabilities of attacker A2.

---

[16] Application Programming Interface.

II.B Attacks on the Communications System

Attacks on the communications system are attacks that lead to threats to privacy or cause drivers to bear extra administrative or legal burdens within the system. These types of attacks occur in two categories:

- **Privacy attacks:** Tracking the location or driving route of a particular person
- **Slander/Framing attacks:** Falsely reporting misbehavior from a vehicle, resulting in an otherwise valid driver being removed from the system

II.B.1 Attack Methods

The attack methods for privacy and framing attacks require a level of technical manipulation of different components of the V2V/V2I system. Three major attack methods include: a) message linking, framing attacks, and Sybil attacks.

In **message linking,** an attacker sniffs V2V Basic Safety Messages, attempting to use information found within messages to identify a particular vehicle or a driver's whereabouts. The Basic Safety Message contains both static[17] and dynamic[18] identifiers, which can be used for message linking. These identifiers help build a dynamic map of the driving environment, which is used to predict vehicle movements, resulting in potential warnings to surrounding vehicles.

Policy Highlight:
Privacy attacks use message linking, meaning that an attacker uses the contents of messages to determine a user's whereabouts. The approach proposes that messages do not contain any personally identifiable information, so that a person cannot be identified by an attacker simply by reading a message.

*In developing the approach, it was assumed that **the approach should not compromise privacy any more than other systems that are in existence today.***

- *Syntactic Linking* – Uses static identifiers to establish multiple messages coming from the same vehicle
- *Semantic Linking* – Uses dynamic identifiers to "join the dots" to reconstruct a vehicle's trajectory using Vehicle Path Prediction methods

A **framing attack** is when an attacker makes a vehicle's on-board equipment appear to be malfunctioning by generating false messages to contradict the target vehicle's legitimate messages. This misbehavior is reported to the CA and could result in an innocent person's vehicle being revoked from the system. This requires an attacker to extract certificates and keys from another vehicle as described in Section II.A on page 19.

A **Sybil attack** is a sophisticated framing attack where an attacker simulates multiple vehicles by using multiple radios, allowing the attacker to send many messages by ultimately posing as several "ghost" vehicles at once. This creates a fictitious environment where it appears as if the targeted vehicle is misbehaving. An attacker needs to obtain valid certificates and keys from the vehicles to perform this attack, thus limiting the number of vehicles affected to the number of certificates obtained. Sybil attacks are a significant risk to privacy because they can be difficult to detect within the system since an attacker is using valid certificates and keys. Also, having

---

[17] Static identifiers include: certificate identifier; source address (at any level); identifier in application message payload; and vehicle characteristics, such as length, weight and vehicle type

[18] Dynamic identifiers include a vehicle's location, velocity, and acceleration.

multiple, simultaneously valid certificates on a vehicle increases the likelihood of a Sybil attack because an attacker can extract these certificates and use them at the same time to pretend to be multiple vehicles at once. The severity of a Sybil attack can be mitigated by restricting the number of concurrently valid certificates for a vehicle. A more in-depth discussion of the certificate management options is included in section III.A.2.3.

II.B.3 Risk Levels

Risk levels for privacy and framing attacks occurring in the V2V/V2I system relate to the appeal of existing methods for carrying out the same type of attack. The risk assessment in Table 2 compares the risk level of tracking a vehicle using the V2V/V2I system versus tracking a vehicle using existing or more traditional methods, such as tracking through cameras, cell phones, and tracking devices. Considering the cost and complexity of launching an attack through the V2V/V2I system, it appears more likely that an attacker would track a vehicle using existing methods, which can be performed at lesser cost and ease. Risk levels also vary between tracking a vehicle at a single location (e.g. person's workplace) and a larger area. An area risk could mean tracking a vehicle's route taken through an extensive area or to monitor overall driving behavior.

The risk assessment in Table 2 assumes certain security measures from the proposed approach are implemented, such as "anti-linking" measures. Anti-linking is a process that incorporates changing all static identifiers included within a message, including certificates, every five minutes, so that these identifiers cannot be linked together to track a person's whereabouts. No PII is contained in any static identifiers, eliminating the possibility that a person's identity can be lifted directly from Safety messages.

Incorporating anti-linking increases the complexity and investment required by an attacker to carry out a successful attack, resulting in lower risk to the system. Implementation of anti-linking significantly reduces the risk level when compared to a scenario where there is no anti-linking. The no anti-linking option is included only to illustrate the change in risk level associated with unmitigated message sniffing (meaning that no efforts are made to prevent linking of Safety messages).

In addition to risk from outside attackers, there is a risk of misbehavior from within the operators and overseers of the system. If the entities responsible for performing security operations were to collude or share information, personally identifiable information could be exposed or become vulnerable. Therefore, it is important that policies regarding data access and enforcement are in place to address individuals or entities wanting to misuse personal data.

Table 2: Assessment of Privacy Risk through Attacks on Communications Infrastructure

	Single Location Risk			Area Risk		
	Preselect Person	Check if preselect person is in data set	Derive person's identity directly from data set	Preselect Person	Check if preselect person is in data set	Derive person's identity directly from data set
TRACKING VIA PROPOSED V2V SYSTEM (Anti-Linking Implemented)						
V2V message sniffing with no personal information used in identifiers & anti-linking	Low	Low	Low	Low	Low	Low
TRACKING VIA V2V SYSTEM (No Anti-Linking Implemented)						
Unmitigated V2V message sniffing	High	High	High	High	High	High
TRACKING VIA EXISTING METHODS						
Cell Phone tracking	Medium	Medium	Low	Medium	Medium	Low
Cameras	High	High	High	Medium	Medium	Medium
Physical tracking	Medium	Low	Low	Low	Low	Low
Tracking Device	High	Low	Low	High	Low	Low
RF Fingerprinting	High	High	Medium	High	High	Medium

Notes:
1) **Preselect person** – an attacker selects a particular person that they want to track
2) **Check if preselected person is in a collected data set** – an attacker could set up a tracking system, and then examine the data set collected from listening stations to determine if the preselected person appeared in the data
3) **Derive a person's identity from collected data** - an attacker may set up a tracking system, and then try to identify a person using only information obtained from within the collected data set

II.C Summary

Expert analysis suggests that the risk and impact to safety is generally low in user attacks and safety impact. The proposed approach makes it necessary for an attacker to have some combination of inside knowledge of the system, physical access to the vehicle, or sizable investment resources to launch a successful attack. Analysis concludes that the greater risk is in reducing acceptance of the system to the point that users deactivate it. With regard to attacks on the system, the main risk appears to be in terms of privacy. The risk to privacy is low as long as anti-linking mechanisms are implemented. An attacker would be more likely to utilize more convenient methods such as cell phone tracking or physically following a vehicle.

The next section describes how the technical and policy elements combine to form a communications security approach that meets V2V and V2I requirements. The description of the approach is further accompanied by a description of how the risks described in this section can be addressed through technical, policy, and a combination of both. The text in section III is annotated with options for system design and configuration.

Section III: Addressing Security Risks—Technical and Policy Options

To achieve a robust communications security solution, the proposed approach combines both technical and policy options to address user and system risks. As described throughout this section, the approach first focuses on the deterrence of attacks through technical design and implementation of enforcement policies. The technical options include design measures for providing security at both the vehicle and system level. If these measures are circumvented, the approach then employs techniques to detect misbehavior and remove the misbehaving entity. As a last measure, where security risks cannot be addressed through technical design alone, policy mechanisms such as governance, legal deterrence, and enforcement can serve a critical supporting role.

III.A Technical Solutions

The proposed approach addresses risks through the prevention and detection of misbehavior. Misbehavior is defined as either a malicious attack (as described in Section II) or a technical defect that occurs through mechanical malfunction. Options for preventing and detecting misbehavior exist at both the vehicle (local) level and through the communications system (global) level.

III.A.1 Mitigating Risk at the Local Level

III.A.1.1 Hardware Security

Hardware security is defined as the **physical security features** that protect the vehicle's on-board unit. Physical security features can protect against attacks that attempt to access the vehicle's on-board unit to extract keys or to manipulate a vehicle's software and sensors[19].

Security experts determined the level of hardware security required on the vehicle by analyzing the risk of successful attacks being launched on users and the system. Several options were compared in terms of cost and the time required for an attacker to compromise the on-board unit. Research indicated that if an attacker had prolonged access to a vehicle, most physical security mechanisms could be circumvented. Therefore, high-security hardware developed specifically for vehicles equipped for connected vehicle communications would not significantly increase the level of

Policy Option:
None of the currently viable commercial hardware options can completely prevent key extraction and other physical security risks once an attacker has gained access to the vehicle. As a countermeasure, legal deterrence and enforcement are needed to compliment the technical measures to protect against equipment tampering.

---

[19] A physical level of hardware security will need to be determined for RSEs and ASDs.

security. A standard controller[20] is proposed for providing hardware security with additional security from software operating on the hardware, rather than through dedicated hardware alone (see section III.A.1.2 on the following page). In addition to the standard controller, it is important to add a level of physical security on the exterior of the on-board unit that makes it difficult for an attacker to extract keys from an on-board unit undetected in a small number of hours. This could be as simple as providing a tamper-evident seal to show unauthorized physical access, similar to gas and electric meters.

Because none of the commercially-available and cost-feasible hardware options can completely prevent key extraction and other physical security risks, it proposed that a system of legal deterrence against physical tampering is used as a countermeasure. This could be similar to deterrence implemented for odometer fraud or tampering with brake lines. Further details on legal deterrence are presented in the next section.

Policy Discussion:
As noted previously, key extraction from the on-board unit poses the most significant risk, with denial of service and denial of communications (jamming) following.

Does the serious nature of these attacks require legal deterrence that breaks with the principles of privacy protections (i.e., implementing the ability to identify misbehaving actors)?

Since this approach may be extended to incorporate other V2I safety, mobility, and environmental applications, it is recommended that safety and non-safety related applications run on separate hardware platforms. Both types of applications would use different controllers but share vehicle infrastructure, such as the radio and vehicle computer[21]. Even with this segregation, attacks on non-safety applications could result in Denial of Service (DoS) attacks for the safety applications if there is shared infrastructure supporting the communications. Additionally, there may be cost implications. Nevertheless, the separation of the processing platform is expected to prevent basic attacks on non-safety applications from compromising the safety applications.

III.A.1.2   Software Security and Misbehavior Detection

The combination of hardware and software implemented at the vehicle level is referred to as the hardware architecture. While hardware security aims to prevent attacks, the addition of software functionality checks and misbehavior detection processes <u>allows for the capability to detect misbehavior that has successfully circumvented physical security measures and entered the system</u>. (Misbehavior is defined as either a malicious attack, such as false messages that have been successfully injected into the system or a technical defect that

---

[20] A Standard Controller and a Standard Controller with Security Features (controller has security features as part of the architecture, i.e. ARM's TrustZone offers virtual processors that cannot interfere with each other) were identical in terms of cost of an attack to the system and the time required for a successful attack to be mounted. The only difference was cost of implementation.

[21] More specifically, the controllers would share the CAN bus. A Can bus is a vehicle bus standard designed to allow microcontrollers and devices to communicate with each other within a vehicle without a host computer.

occurs through mechanical malfunction[22].) Implementing this capability is especially important for detecting attacks that cannot be mitigated through physical security, such as Denial of Service attacks which are launched remotely.

If an attacker successfully circumvents the physical security measures on a vehicle, several software functionality checks running on the hardware will detect equipment tampering. Software checks at the vehicle level include:

- **Secure flashing**: is a process that protects the software running on an Electronic Control Unit (ECU)[23], by ensuring that only authorized software can run on the ECU. In this case, the ECU is the on-board unit. Secure flashing combats software manipulation attacks by detecting software that was installed by an attacker designed to create false messages, alter sensor input information or extract keying material. This also ensures that the core functionality of the DSRC radio cannot be compromised.

- **Component identification:** is a process used by the ECUs in a vehicle to identify each other. ECUs establish a "home environment" and if an ECU is taken from one vehicle and put into another vehicle, it stops functioning. It does not prevent attacks but adds another layer of difficulty the attacker has to overcome. By allowing the DSRC radio to participate in the vehicle's component identification scheme, this would prevent the DSRC radio from functioning outside the intended vehicle, mitigating attacks where the radio is stolen or removed from a junked vehicle and directly installed into another vehicle.

- **Plausibility checks**: detect bogus sensor data generated from either manipulated or faulty sensors on both the sender and receiver vehicles. These checks run continuously on all messages between vehicles. If the vehicle can detect that the message received is bogus or spurious[24], then it will not relay this message to other vehicles. It is expected that almost all technical defects would be able to be detected by the on-board diagnostics of the sending vehicle and defective messages would not be sent out. One way that a vehicle checks plausibility is by building a physical model of its surroundings and checking it for consistency – known as location checks. Consistency checks can also be internal to ensure that the current state of the on-board unit is consistent with the last known state. Denial of Communication attacks cannot be prevented through physical security but can be detected through plausibility checks.

III.A.1.3 *Misbehavior Detection at the Local Level*

Random message checking is a process that collects messages from vehicles to be sent to the certificate authority for further checking. First, local processing checks the validity of each message through plausibility

---

[22] Software would identify both as misbehavior because the system cannot determine the cause of a defect. The failure rate assumption used in the security research considerations was 1 failure in 1,000,000 hours of operation. Almost all technical defects would be able to be detected by the on-board diagnostics, and consequently a vehicle would be able to avoid sending out messages that would be judged to be misbehavior due to a technical defect. In combination with plausibility testing, the likelihood of a technical defect without the sender noticing it is extremely small. Precise numbers are not available.

[23] An ECU is a generic term for any embedded system that controls one or more of the electrical systems or subsystems in a motor vehicle. Many vehicles have over 70 ECUs.

[24] Bogus messages use an invalid signature, and spurious messages include a valid signature but incorrect payload data.

checks at the vehicle level. Local processing is performed without active interaction with other vehicles or with the certifying authority. After verifying incoming messages, the vehicle then collects a random number of messages (referred to as reports) to be sent to the CA for further checking, known as global processing. Reports could include a record of an event as well as random non-suspicious messages. This builds a global view for detecting misbehavior. This is an important tool for detecting Sybil attacks since the CA is collecting reports from many vehicles at once and can detect that certificates apparently used by different vehicles were all in fact extracted from the same OBU (using authority traceability). This process of misbehavior detection allows for certificate revocation of bad actors, due to both technical defects and malicious attacks. More discussion of global processing and authority traceability can be found in section III.A.2.5 on page 34.

> *Policy Discussion:*
> *To detect misbehavior and appropriately revoke a user from the system, the certifying authority must use "authority traceability" to identify the misbehaving entity. A decision must be made regarding how far back in time messages can be linked to a vehicle by the authority.*
>
> *Depending upon how the authority decides to enforce misbehavior, either through simple revocation from the system or other legal options, this could reveal a person's identity and impact privacy.*

It is expected that some false messages will manage to circumvent the software and misbehavior checks. Using the proposed configuration, estimates are that a user will encounter a false message once every four days. Although there has not been extensive research conducted on how drivers will react to a false message, initial research suggests that this type of error does not pose a significant risk to safety. The driver is ultimately in control of the vehicle and can ascertain (in most situations) whether it makes sense to react to a message. The greater risk of false messaging is that it could ultimately reduce driver acceptance of the system. Drivers could choose to ignore accurate messages because of the assumption that some messages are providing false information. This behavior may pose a safety problem in that a driver may ignore an accurate message that could have prevented a collision.

> *Policy Discussion:*
> *Is the **encounter with a false message once every four days** an acceptable level for public acceptance of the system if the level of risk to safety is low?*

Suppressed messaging is the result of Denial of Service (DoS) attacks and refers to instances when a warning should have been issued in a scenario (e.g. crash avoidance application), but did not occur. The near-term consequences of suppressed messaging can cause an outage of collision avoidance warnings, which returns drivers to the level of awareness that existed when there was no system. However, these risks may increase over time as users become accustomed to relying upon generated warnings and thus are likely to place less weight on his/her own evaluation of an imminent collision within the surroundings. To counteract this risk, if an attack is detected the approach suggests that the receiving unit issue a "Service Temporarily Unavailable" message to protect a driver from getting confused between the absence of a message due to no danger, and the absence of a message due to channel congestion resulting from a DoS attack.

III.A.2 Mitigating Risk at the Global Level

Protecting privacy at the Global level includes options for implementing "anonymous, randomly-assigned identifiers with each message" and further ensuring that messages are "unlinkable". Anonymous identifiers mean that a user cannot be identified through content included in the messages. A user is said to have unlinkability when it cannot be determined that two messages belong to the same user.

III.A.2.1 Public Key Infrastructure

A Public Key Infrastructure (PKI) to establish trust within the communications network. Its purpose is twofold: to provide authorization credentials to users for participation in the network and to revoke those credentials if an administrating authority decides to do so. The PKI design incorporates the use of authorization credentials to support trusted communications between users. Authorization credentials in a PKI are known as certificates, which represent information cryptographically bound together and certified by an administrating authority called a certificate authority (CA). Although this approach is presented in terms of vehicles communicating with the certificate authority, it is proposed to be applicable to other mobile users[25] during full deployment.

III.A.2.2 Certificate Authority

An important component of the approach to security is a Certificate Authority. A CA is an authorizing organization that issues and revokes certificates, and maintains and distributes certificate status information. It is generally a trusted party that is central to the security services of the system. A CA can detect misbehavior occurring within the system as a result of technical malfunction or intentional malicious attack. In either case, the CA can revoke the certificates associated with this misbehavior and inform other users within the connected vehicle environment that these revoked certificates are no longer trusted.

Communications between vehicles and the CA must be trusted and secure. "Secure two-way" in this context means establishing a communications session with trusted credentials at each end that use encryption of messages sent over wireless communication and backhaul technologies to prevent eavesdropping and tampering. In the proposed communications exchange (see Figure 2 on Page 15**Error! Bookmark not defined.**), the CA issues certificates, private keys, and information about revoked certificates to the vehicle through a communications infrastructure node or existing network, allowing the vehicle to obtain new certificates and keys, protect itself against bad actors, and ensure that communications with other vehicles are trusted.

The expert teams evaluated several communication scenarios that take into account the amount of bandwidth and access time required for communication with the CA. Greater available bandwidth and lower time-to-access provides for a more effective the security mechanism. However, cost and privacy considerations prohibit continuous connection of vehicles to the CA.

---

[25] Mobile users include any user device, such as a vehicle device (as described in this approach), pedestrian smartphones, etc.

Proposed Organizational Structure for Certificate Authority

There is a privacy risk associated with the CA, stemming from the fear that the CA itself could be untrustworthy or act maliciously. The proposed approach recommends incorporating a "split CA", which divides the certificate authority's capability and functionality by both technical and organizational means in order to enhance privacy. Splitting the CA creates a situation where one administrating authority does not hold enough personally identifiable information to compromise a person's privacy or vehicle's identity.

As a hypothetical example, the overall CA structure can be divided into a Request and Registration Authority (RA) and a Certificate Authority (CA). The RA verifies requests for keys and certificates and authorizes assignment of keys and certificates. It determines whether a certificate request should be granted, but will not be involved in creating and assigning the keys and certificates to vehicles or other mobile users. The CA creates, stores, and assigns keys and certificates to vehicles authorized by the RA. The CA also is responsible for credentials management, CRL management, and receiving misbehavior detection reports. The CA recognizes invalid certificates from the misbehavior detection agent, adds the vehicle revocation identifier on the CRL, signs the CRL, and performs all functions of key management.

Splitting functions between entities could have cost implications, although the expert teams indicated that this cost would be minimal and without impact to the communication channel requirements. There is a research effort being launched to develop organizational and operational models (options) for the certificate authority that will provide greater detail on costs and effectiveness associated with splitting the CA.

> *Policy Discussion:*
> *A CA insider or the CA itself could be untrustworthy or act maliciously. This risk is addressed by dividing responsibilities and information across multiple CAs or between a CA and a Registry Authority (RA). No one authority would hold enough information to link a certificate or message to a vehicle identifier or actual driver identity.*
>
> *The segregation of information across multiples entities carries cost implications. Analysis to date suggests that the institutional costs are likely to be small by comparison to the gains in privacy protection and public acceptance. The policy research effort to develop organizational and operational models (options) will provide greater detail on costs and effectiveness associated with splitting the CA into multiple entities (See Appendix A).*

> *Policy Discussion:*
> *User safety and misbehavior detection may require the CA to maintain short-term linkability with the vehicle. Further stakeholder discussion is needed to determine the appropriate balance among these safety requirements with the importance of privacy.*

III.A.2.3 Certificates

Certificates are necessary for users to verify that other users are authorized to use the network and can therefore be trusted. To protect user privacy, certificates do not contain identifying information, but instead use random pseudonyms as temporary identifiers in messaging. However, even with this measure in place, using a single certificate over a large time-span would allow long-term tracking of a vehicle. Therefore, the approach proposes that each vehicle is loaded with a set of certificates that are changed regularly. Privacy is enhanced with more frequent changes but this also increase local storage and communication requirements.

Proposed Certificate Management Approach

Several options for the certificate were identified with the intent to promote maximum safety benefits, but to also protect user privacy by supporting anonymity and unlinkability.

Each vehicle is equipped with two different types of certificates:

- Short-Lived Certificates (SLCs) – These certificates are used during normal mode of operation and are changed every five-minutes.
- Long-Term Certificates (LTCs) – These certificates are used during fail-safe mode during low-likelihood events when SLCs may not be available.

The following is the proposed certificate protocol:

- Each SLC is only valid for a single, predetermined, time-restricted five-minute period. Each certificate will have a fixed start and end time and is valid only during the specified period, regardless of whether it is actually used during that time or not.
- All vehicles share the same time restricted intervals so an attacker is not able to distinguish between any two vehicles based on its start and end times. As an example, the CA issues all certificates so that the first certificate is valid on January 1, 00:00:00 – 00:05:00, 00:04:30 – 00:09:30, and so on. Fixed times also increase the protection against Sybil attacks by limiting the number of certificates available at any given time.
- SLCs will overlap with preceding and following certificates for 30 seconds to ensure that the vehicle always has access to a valid certificate. It is believed that a 30 second overlap provides enough time to avoid any outages due to failed certificate changes, as well as ensuring communications during critical circumstances, such as a certificate change delay during a pre-crash safety communication.
- All static identifiers[26] contained in the Basic Safety Message, including the certificate's temporary identifier, will be changed simultaneously every five minutes. Changing certificates and other identifiers often is important to protecting privacy because it makes tracking a vehicle via information contained in unencrypted Basic Safety Message (BSM) more difficult.
- A certificate's temporary identifier field within the Basic Safety Message (BSM) will be randomized. Other static identifiers, such as vehicle length and width, will not be randomized, but instead will be specified loosely enough that a particular vehicle cannot be identified by these types of static fields. [27]

---

[26] Static identifiers include: certificate; source address (at any level); identifier in application message payload; fields that may appear in the BSM (such as vehicle data).

[27] An attacker may also attempt to use the vehicle's Basic Safety Message (BSM) to track a vehicle. The BSM contains static identifiers – such as the vehicle's temporary identifier (in place of using PII) and a vehicle's physical characteristics. These properties assist to build a dynamic state map of the driving environment, which is used to predict the future movements of the vehicle, resulting in potential warnings to surrounding vehicles. More research will be conducted during the Safety Pilot to determine the trade-offs between breaking an attacker's ability to perform linking (by not including fields such as precise length, weight or vehicle type data), and the safety risks posed by breaking a vehicle's ability to building an accurate dynamic state map.

The five-minute validity recommendation was established based upon discussion among the security experts when evaluating the trade-off between privacy, safety, and security. A 10-minute validity gives the attacker a likelihood of almost 28% to compromise privacy. A five-minute validity gives the attacker a likelihood of around 7% to compromise privacy. A two-minute validity essentially avoids privacy compromise. However, a two-minute validity could begin to have a more significant negative impact on safety since an identity change temporarily weakens a vehicle's ability to track surrounding vehicles. Furthermore, switching certificates below the five-minute threshold could compel an attacker to use less expensive and technologically advanced methods of tracking, such as radio frequency fingerprinting. Therefore, it was deemed that the five-minute certificate validity period to be a feasible approach. This is a technical recommendation but does require agreement from privacy advocates that this is an appropriate approach.

Policy Discussion:
*The approach includes changing certificates every five minutes. This time period was selected based on research that estimated a **7% chance of an attacker compromising privacy**. A validity period of two minutes essentially avoids privacy compromise, but presents a higher cost associated with certificate storage and communication with the CA.*

The proposed approach also addressed instances when short-lived certificates may not be available due to time restricted certificates expiring. An example is when a user is on vacation for an extended period of time and has not been in communication with the CA and is therefore unable to unlock the next batch of short-lived certificates. Vehicles will go into fail-safe mode and use LTCs in these cases. The availability of long term certificates assures uninterrupted participation in the system. Thus, upon return to driving, the vehicle's ability to use safety applications will not be compromised due to an expired certificate and safety applications will function until the vehicle unlocks new SLCs. This has a temporary impact on privacy since a vehicle could be tracked for longer periods through the use of the LTC. It has not yet been determined whether one or more long-term fail-safe certificates per vehicle will be required; or the validity time period for these certificates. The availability of long term certificates also increases the number of available certificates to an attacker, thus increasing the likelihood of a successful Sybil attack and risk to privacy.

III.A.2.4 Certificate Management and Distribution

The manner in which the CA issues and revokes certificates is defined as the Certificate Management scheme. The proposed approach uses Linked Certificates – meaning that certificates will be cryptographically linked together by an encrypted identifier. The certificates are stored on the vehicle and are encrypted until unlocked by a decryption key. The vehicle's on-board unit must request this key from the CA. Each key request would unlock a different group of certificates. This reduces the possibility that a bad actor can gain access to a large number of certificates at one time, thus reducing the likelihood of a Sybil attack.

Since the use of time-restricted certificates introduces the need for a greater number of certificates per vehicle, all certificates will be encrypted and loaded onto the vehicle at longer-term intervals, for example, annually. The certificates are stored in batches, or bundles, and cannot be unencrypted until a key is received from the CA. At full deployment, vehicles will request keys from the CA daily to unlock certificates. Vehicles will receive certificates, keys and CRLs through infrastructure nodes, most likely deployed as roadside units (RSEs).

Revocation information will be distributed in the form of a certificate revocation list (CRL) by the CA. Each revoked vehicle requires only a single CRL entry containing the small revocation key since all certificates are linked, as compared to putting all individual revoked certificates on a CRL. This decreases the size of the CRL significantly, reducing the bandwidth required for CRL communications. The CA can then deny requests for new certificates or decryption keys based on this list. This recommendation allows vehicles to be informed sooner, but requires more communication (in terms of bandwidth) between vehicles and the CA for aspects of certificate management other than CRL distribution.

III.A.2.5 Misbehavior Detection at the Global Level

Options for misbehavior detection exist at the local level and at the global level. As discussed earlier, some attacks can be detected at the vehicle (local) level through software functionality checks and other misbehavior detection processes. While local level processes result in a vehicle not accepting false messages, it does not address the removal of the misbehaving actor from the system. This is a global level misbehavior detection process. Global processing provides a method for gathering a system-wide view of misbehavior to detect attacks that may not be geographically limited. Global processing incorporates random message checking to collect a certain number of local messages generated by the vehicle (referred to as reports) to be sent to the CA for further checking. Misbehavior reports are collected from many vehicles and compiled to determine whether a reported vehicle is misbehaving through analysis of message content, sensor data, onboard dynamic state map, or physical laws. Since the CA is collecting reports from many vehicles at once, the CA can more easily detect Sybil attacks (as compared to local processing detection) by detecting that certificates apparently used by different vehicles were all in fact extracted from the same OBU.

Policy Discussion:
Frequent vehicle reports to identify misbehavior increases security by allowing the authority to quickly discover bad actors within the system. The trade-off may be increased risk to privacy, as linkability is required to take action on a misbehaving vehicle.

Also, collection of more frequent reports is a trade-off between increased security and more CA communication which may increase congestion on the channel as well as increase CA administration costs.

Options surrounding what information will be included in the vehicle reports, how often reports are collected, and what constitutes misbehavior will be included in the analysis for organizational and operational models.

In order for the CA to revoke misbehaving vehicles from the system, linkability is needed to allow for messages sent by a vehicle to be linked to the vehicle's cryptographic identity. This is known as <u>authority traceability</u>. The authority traceability property makes the reporting and the reported vehicles accountable to the CA. The actual decision of whether the reported vehicle is malicious or a victim of a slander attack would be made by the CA after examining all of the available contextual information. Authority traceability may increase the risk of an insider tracing a particular vehicle, but this risk can be mitigated through the split CA functions.

III.A.2.6 Communication Network – Bandwidth and Access Time Requirements

With an understanding of the risks and mitigation features described throughout section III, the remaining analysis focused on the frequency of a vehicle's communication with a CA to update certificates, the revocation list and to report misbehavior. While frequent communications increases security, there is a direct trade-off between access time for vehicle and CA communications and bandwidth usage.

The analysis did not assume a specific communications media when developing the requirements for access time and bandwidth. Instead, it looked at the amount of bandwidth required for security credentials management based on the potential vehicle population (assumed at 250 million vehicles as the basis of a nationwide deployment).

Certificates linked to misbehavior should be revoked quickly to reduce the window of opportunity for misbehavior to damage the network. It is assumed that there will be fewer instances of attacks with less deployment. Three levels of deployment were considered in establishing the Time-To-Access (TTA)[28] and the time it takes to revoke an attacker—10 percent deployment, 50 percent deployment, and 100 percent deployment. The results are in the table below:

Level of Deployment	TTA	Time between the start of an attack and the revocation of an attacker
10%	Every 10 days	25 days
50%	Every 2 days	5 days
100%	Once every day	2.5 day

With this analysis, bandwidth requirements and the ideal options for how and where a vehicle communicates with a CA (for instance, on highways, local arterials, at gas stations, etc.) can be determined.

> *Policy Research:*
> *Research is needed to establish criteria for an effective process for end-of-life for on-board equipment. Long-term certificates (LTCs) could be valid for a vehicle's lifetime. Further policy research is needed on the implications of "junking" a vehicle or transferring ownership. If the LTCs are not revoked at the end of the vehicle's life, this opens the possibility of attackers stealing certificates from junked vehicles. Such issues will be explored further in policy research that seeks to identify legal enforcement options.*

---

[28] The time between two consecutive times a vehicle communicates with a CA is called Time-To-Access (TTA).

Section IV. Preliminary Policy Analysis

As noted in the side textboxes throughout Sections I-III, technical design options may rely upon policy solutions to provide comprehensive security. This section identifies the policy and institutional elements and describes the type of policy research being conducted in support of the approach. The text additionally notes some of the inherent conflicts that may require decision makers and stakeholders to balance priorities.

The policy and institutional issues that are considered most significant (and thus will result in additional policy research) include:

- Analysis on how to most effectively design organizational and operational entities that will support security credential (certificate) management, legal deterrence, misbehavior detection, and revocation. Policy and institutional issues include questions on cost, whether to split the entities for enhanced privacy, personnel and equipment needs, and policies and procedures.

- Identification of the specific types of legal deterrence and policies that will act to prevent or mitigate misbehavior within the system. Policy questions include the determination of authority for enforcement.

- Development of a strategy for updating and implementing the 2007 privacy principles, including development of practicable options for putting the principles into use.

- Analysis on implementation options that compares different configurations using infrastructure and non-infrastructure options. Further, analysis on the types of sustainable funding, financing, investment, and/or revenue sources available with these implementation options that address the needs for funding initial deployment as well as ongoing operations, and maintenance.

- The identification of the level and type of governance and authorities required for implementation of the organizational and operational models and with the communications data delivery system.

IV.A Organizational and Operational Models for a Certificate Management Entity

A technical solution for greater privacy protection involves the use of a split CA. Splitting the CA creates a situation where one administrating authority does not hold enough personally identifiable information to compromise a person's privacy or vehicle's identity. By splitting the CA functions, multiple organizations must intentionally collude and act together to negatively impact personal privacy.

In addition to assessing the options for how the organizational and operational entities might be configured, further policy research is needed to address:
- The appropriate level of institutional resources that are needed and whether existing institutional arrangements can be leveraged. Included is an analysis of the public/private sector roles and determination of whether the entities must be wholly public, can be wholly private, or require some level or partnership;

- The costs associated with security credentials management as well as organizational and operational costs; and
- The governance authority needed to provide legal enforcement.

In determining how to structure a CA for V2V/V2I, policy research is underway and will result in the development of options for:

- Configuration of a certificate management entity(ies) and how they might split functions across multiple authorities. The options will be accompanied by proposed policies for enforcement and for overseeing the separation, and definitions for clear roles and responsibilities within the organizational structure(s).
- Policies and procedures regarding misbehavior—whether to simply remove the misbehaving actor or to further link the actor to a personal identity in order to legally address the misbehavior and/or revoke the privileges of certain users or vehicles. While linking the action to a person provides a more complete security approach, it violates the principle of anonymity. Policy decisions must still be made on how this policy option should be implemented, especially in relation to the overall security context.
- Governance options and analysis of public and private roles and/or partnerships.

Further detail on this policy research is provided in Appendix A.1.

IV.B Legal Deterrence and Enforcement

The proposed security approach must assume some level of legal deterrence and enforcement to address potential bad actors. As shown in other industries, strict legal consequences paired with effective enforcement methods can deter bad actors and strengthen the overall security approach.

For V2V/V2I communications, legal deterrence is especially important for limiting improper physical access to vehicles. Limiting physical access is essential to a vehicle's hardware and software security. Similar to how illegal access and tampering to personal vehicles are currently addressed through legal enforcement, a comparable policy technique could be used for vehicles equipped with connected vehicle safety technologies and applications.

Legal deterrence and policies for enforcement are also critical to addressing misbehavior. The current design proposes that misbehavior can be detected, resulting in the bad actor being removed from the system (through revocation of certificates). An additional option would be to identify the misbehaving party(ies) in order to take further legal action. This would require cooperation between the split CAs for identification, but would also violate the privacy principles as they stand today. Included with the research on certificate management entities (Appendix A.1) is research to identify which actions would be considered illegal and the types of enforcement techniques available within the limitations of the system design and the privacy principles. A separate legal analysis will be performed to classify the legal and illegal actions and to determine how best to address them from a user perspective while considering concerns about privacy and crime.

IV.C Privacy

As defined today, the security approach provides three important conclusions regarding privacy when implemented in support of vehicle-to-vehicle communications for crash-avoidance safety applications. They include:

1. No personally-identifiable information (PII) is available in the system and thus, an attack on the operational system cannot violate privacy
2. Short duration of tracking is possible but the attack cannot obtain vehicle identity through the system and the security approach makes it *very difficult for this to occur (requires complex equipment which means a sizable investment)*
3. Vehicle identification is available only through integration of information from each certificate authority, *a risk that will be addressed through policies on access to systems, hiring/ employment checks, and proper enforcement in place*

The above conclusions will need to be revisited from the perspective of mobility, environment, and convenience applications that may require a user to identify a device (i.e., when requesting routing information) or provide other information (such as financial information for parking). Other remaining issues that still require analysis are:

- Given the requirement of anonymity by design, what are the implications for cost, communications load, and deployment?
- Given the requirement of anonymity by design and the privacy principles, would it ever be appropriate to identify an individual as a bad actor? Is this trade-off acceptable to the public in terms of using and trusting the system?
- How effective are existing policies on system access, hiring/employment, and improper use of data, based on lessons learned from other industries? (will be addressed through research described in Appendix A.1)
- Is this approach to privacy acceptable to privacy advocates, safety advocates, and the public?

IV.D System Implementation: Costs and Sustainable Funding/Financing/Investments and Levels of Security

Identifying the ability to finance system implementation and provide **sustainable funding for operations and maintenance** is another critical policy issue under analysis. The level of system installation and ongoing costs may vary significantly based upon the type of deployment or implementation model used. For instance, a system model that requires new infrastructure is likely to cost much more at implementation than one that leverages existing infrastructure or other equipment (for instance, cellular or Wi-Fi networks).

There are three important elements to considering costs—costs for developing the security approach, costs associated with implementation, and the ongoing cost of operations and maintenance (importantly, in terms of operations, most communications security systems are not designed and implemented on the scale envisioned for the V2V/V2I system). An important element to consider when identifying options for financing, partnerships or investments is whether options exist to create a revenue stream that would help sustain ongoing costs. Thus, further research into viable models is needed. Appendix A.2 describes the research efforts associated with these requirements.

IV.E Governance

Technical solutions play an important role in mitigating risk, deterring misbehavior, and protecting privacy. Policy and institutional solutions result in institutional processes for organizations, addressing user access, rules of operations, enforcement and other important elements of a robust operational system. Policy research is important for determining whether the implementation and ongoing operations of a secure communications data delivery system is viable.

Eventually a set of decisions will be needed to determine who will take on the roles and responsibilities for implementation, operations, oversight, maintenance, and conflict resolution. This question falls under the issue of governance. Other governance questions include, but are not limited to:

- Can governance (as well as the institutional solutions for CA and security) be wholly private, or must it be wholly public, or some mix, and what is an appropriate balance within that mix?
- Who will implement the systems and solutions?
- Who will provide oversight? Decision making? Who will finance what part or all of the system?
- Are any new authorities needed with respect to enforcement, rules of operation, or other decisions?

Research into governance options will be informed by the technical and policy research described in this document. An initial step was taken with the development and hosting of a Governance Roundtable on June 20, 2011. Experts from other industries were invited to discuss the steps in analyzing what governance is needed with the introduction of new technologies and systems. Experts provided insight into existing models that could inform options for governance for the connected vehicle environment. Proceedings from this event are posted on the ITS Program website and form the basis for developing a strategy for next steps.

Section V. Conclusion

This document presents the options that form the basis for an approach to V2V/V2I communications security. The options were derived from an analysis of risks and existing industry best practices. The proposed approach is a combination of technical and policy options that results in a trusted and secure system for V2V/V2I communications and users, and one that adheres to specific conditions for privacy and scalability. The proposed approach reasonably meets these objectives and further considers the balance required in prioritizing safety and security requirements with costs, privacy, and institutional and governance requirements.

In summary, it is possible to combine hardware, software, and policy measures into an approach that can mitigate high impact attacks, making attacks highly improbable given cost and level of effort required for a meaningful attack, and thus assuring acceptance by users for trusting the system and for using applications for crash-avoidance safety. The known advantages of the approach are that it:

- Meets the objectives of providing trusted, anonymous messages using random identifiers that are changed every five minutes
- Is scalable to 250+ million users
- Supports crash avoidance safety applications
- Many attacks are only feasible with a significant amount of investment and expertise about the system
- Approach prevents/mitigates against harm to the system.

Some of the known limitations that will be addressed through further policy research to examine options include:

- There is no instantaneous identification of misbehaving actors. Inevitably, there is a delay in identification and delay in removing misbehaving actors from system.
- Splitting the certificate management entity may have cost implications.
- Currently, there are no clear financial models that offer a sustainable foundation for ongoing operational and maintenance costs.

Appendix A: Details on Policy Research

A.1 Policy Research for Certificate Management Entities:

The research into certificate management entity(ies) will result a review of best practices and lessons learned as a basis for tailoring an entity to meet V2V/V2I requirements:

- An evaluation of whether a Certificate Authority is the most appropriate type of entity to provide the communications security, certificate management responsibilities or whether there are other types of certificate management entities for consideration. This will include evaluation of relevant best practices and a comparison of entity types, discussing advantages and disadvantages of each.

- The challenges and risks to developing an operational/organizational model for a CM Entity under the existing Communications Security approach.

- Policy issues and trade-offs to consider when designing a CM Entity and a certificate management scheme (e.g. privacy, security, institutional issues, governance, and other key topics).

- An investigation into investigating varying institutional designs, ownership structures, decision-making processes and policies. This research will focus on the benefits, costs, and risks associated with wholly public, wholly private, quasi-public, or public-private partnership structures and evaluate how these varying types of organizations would be able to meet identified objectives of the communications security approach. It further includes a discussion of the Federal and/or government role in each of the organizational types.

The initial review will provide a basis for developing organizational structures and options from a physical operations perspective (i.e., centralized versus decentralized, or regional); from a cost perspective and ease of implementation (including impacts on the public sector and on industry), and from a cross-border perspective. Roles and responsibilities will be defined and will include (but not limited to) options for who is suited to:

- Manage user access and certificate issuance and to detect misbehavior and revoke certificates or distribute certificate revocation information;
- Store and retrieve data for misbehavior detection purposes and to backup databases ;
- Perform or provide security – physical and logical access;
- Establish and monitor system performance metrics and audit policies and procedures;
- Manage user privacy protection and provide enforcement; and
- Provide system administration and maintenance.

The models that result from the research will include proposed rules of operation/use, standard protocols, operational and decision making policies. These rules and policies are expected to include (but are not limited to) options for:

- Standard operating rules and protocols;
- Procedures for certificate issuance – how to register users; generate, sign, encrypt, and manage certificates; and manage revocation functions and information;
- Procedures for identifying, processing, and handling certification revocation and enforcement options for addressing and deterring misbehavior, including legal limitations of enforcement, and options for addressing poor system performance or failing devices;

- Decision-making processes and roles for making continuous improvements, changes, and updates to the rules of use, standard protocols, and operational and decision making policies;
- Backup plans and identification of disaster recovery levels required by functional objectives; and
- An analysis of how rules of operation/use are impacted when operating nationally, i.e., an entity operating across multiple states may face specific multijurisdictional and other policy issues.

Policy research will also provide options for rules of access to the certificate management system and proposed policies for all system participants: users, general public, system administrators, entity staff, deployment agencies, and the U.S. DOT. Rules of access are expected to include (but are not limited to) proposed options for:
- Processes for data administrators to securely access and manage data within the certificate management system and policies and processes for obtaining access for different system participants
- Policies that protect the system from unauthorized users; and
- Policies and processes for specific types of requests for access (e.g. law enforcement requests, requests from government agencies) and analysis of whether and how such requests might conflict with the privacy principles.

Ultimately, the policy research will result in an ability to construct full-scale organizational and operational models for real-world operating conditions. These models will include, but are not limited to, analysis of:
- Resources requirements including staffing by functional skill sets; facilities; equipment; and operating requirements.
- Costs including staffing, facilities, equipment (hardware, software, etc.), operations (daily, annual costs, etc.), maintenance (of equipment, etc.), administrative costs, other ongoing/recurring costs, and overall implementation costs.
- Implementation requirements which are expected to include an assessment of implementation steps and timelines (short and long-term milestones); an identification of task owners (roles/responsibilities) in implementation; resource requirements (level of resources needed and associated timelines/costs); authority/legal requirements (if necessary); and risks, challenges, and institutional barriers to implementation.

A.2   Policy Research for Communications Data Delivery System Options:

Research that is planned or underway includes development of **options and business models for the communications data delivery system** for which communications security is required. Research includes an analysis of:
- The effectiveness and cost-effectiveness of various infrastructure and non-infrastructure configuration options for meeting requirements, particularly the requirement for frequency of security credential materials distribution;

- The capability of meeting key stakeholder requirements, particularly the security, reliability, and privacy required when distributing and managing credentials;
- The ability to meet performance metrics for coverage, reliability, redundancy, frequency of security credential updates, and ability to identify misbehavior, at a minimum;
- The acceptability of each option with respect to security; privacy principles; Federal, State, regional, local, and cross-border laws and regulations;
- The financial sustainability of each option, including the ability to leverage existing systems; the mechanisms that offer sustainable revenues in support of lifecycle costs (capital, operating, and maintenance); and the opportunity to attract investment partners or revenue sources;
- The challenges and opportunities associated with extending a communications delivery system primarily focused on communications for security credentials distribution and management to provide support for other applications including V2I safety, mobility and/or other uses of the connected vehicle environment; and
- The ability to implement and sustain the business model.